Math in Our SOLAR SYSTEM

Applying Problem-Solving Strategies

Barbara Linde

PowerMath™

The Rosen Publishing Group's
PowerKids Press™
New York

Published in 2005 by The Rosen Publishing Group, Inc.
29 East 21st Street, New York, NY 10010

Book Design: Daniel Hosek

Photo Credits: Cover, pp. 3 (background), 4 (background), 6 (background), 8 (background), 11 (background), 12 (background), 13, 14, 15 (background), 16–17 (background), 18, 19 (background), 20 (background), 22–23 (background), 24–25, 26 (background), 28 (background), 30 (background), 31 (background), 32 (background) © Photodisc; pp. 5, 7, 10, 18, 19, 21, 23 (planets) © Digital Vision; p. 9 © Roger Russmeyer/Corbis; pp. 16–17 (Mars), 29 (Stardust) courtesy NASA; p. 27 © Bettmann/Corbis; p. 29 (Hubble) © NASA/Corbis; p. 29 (International Space Station) © Reuters/Corbis.

Library of Congress Cataloging-in-Publication Data

Linde, Barbara M.
Math in our solar system : applying problem-solving strategies / Barbara Linde.
p. cm. — (PowerMath)
Includes index.
ISBN 1-4042-2936-1 (library binding)
ISBN 1-4042-5135-9 (pbk.)
6-pack ISBN 1-4042-5136-7
1. Mathematics—Juvenile literature. 2. Problem solving—Juvenile literature. 3. Solar system—Juvenile literature. I. Title. II. Series.
QA40.5.L56 2005
510—dc22

2004003351

Manufactured in the United States of America

Contents

What's in the Solar System? 4
A Huge, Hot Star 6
The Inner Solar System 8
The Outer Solar System 18
Small Space Objects 26
Man-Made Space Objects 28
Space Strategies 30
Glossary 31
Index 32

What's in the Solar System?

When we look at our solar system, we can start with the sun. The sun is the center of our solar system. Nine planets travel around the sun in their orbits, or paths. Mercury is the closest planet to the sun. Venus, Earth, and Mars are next. Farther away from the sun are Jupiter, Saturn, Uranus, Neptune, and Pluto. Most of the planets have moons that orbit them. So far, we know of more than 135 moons, but there may be more. The solar system also contains comets, meteors, and **asteroids**. Since the middle of the twentieth century, new objects have been added to the solar system. Man-made **satellites**, telescopes, **probes**, and spacecraft are also in orbit in the solar system.

The solar system is a big place, and there are many mathematical questions we can ask about it. When you need to find the answer to a math question, there are different **strategies** you can use. Sometimes it is just a matter of solving a mathematical equation using subtraction, addition, multiplication, or division. Other times, it may be helpful to make a chart or table, look for patterns, or estimate an answer.

This picture shows the sun and the 9 planets that orbit it. The actual planets are millions of miles away from each other. Earth, for instance, is 93,000,000 miles away from the sun!

Mercury
Venus
sun
Earth
Mars
Jupiter
Saturn
Pluto
Neptune
Uranus

A Huge, Hot Star

The sun is a huge, hot star. Life on Earth needs the sun's light and warmth to survive. The sun has a **nuclear** furnace in its core. Very high pressure and temperatures in the sun's core change **hydrogen** gas to **helium** gas. This change produces large amounts of energy. The energy leaves the sun in the form of heat and light. Planets closer to the sun get more of the heat and light. Planets farther away get less heat and light. Earth gets the perfect amount of heat and light to sustain life.

The temperature of the sun's core is about 27 million degrees Fahrenheit, or 27,000,000°F. The sun's surface temperature is about 10,000°F. How can you find out the difference between the temperature of the sun's center and the temperature at its surface? We need to analyze the question to choose the best strategy. Which of the operations shown here is the correct one to use to find the difference between the 2 temperatures?

subtraction

$$\begin{array}{r} 27{,}000{,}000^\circ F \\ -\quad 10{,}000^\circ F \\ \hline 26{,}990{,}000^\circ F \end{array}$$

addition

$$\begin{array}{r} 27{,}000{,}000^\circ F \\ +\quad 10{,}000^\circ F \\ \hline 27{,}010{,}000^\circ F \end{array}$$

To find the difference between 2 numbers, we need to subtract the smaller number from the larger number. The center of the sun is 26,990,000°F hotter than the sun's surface!

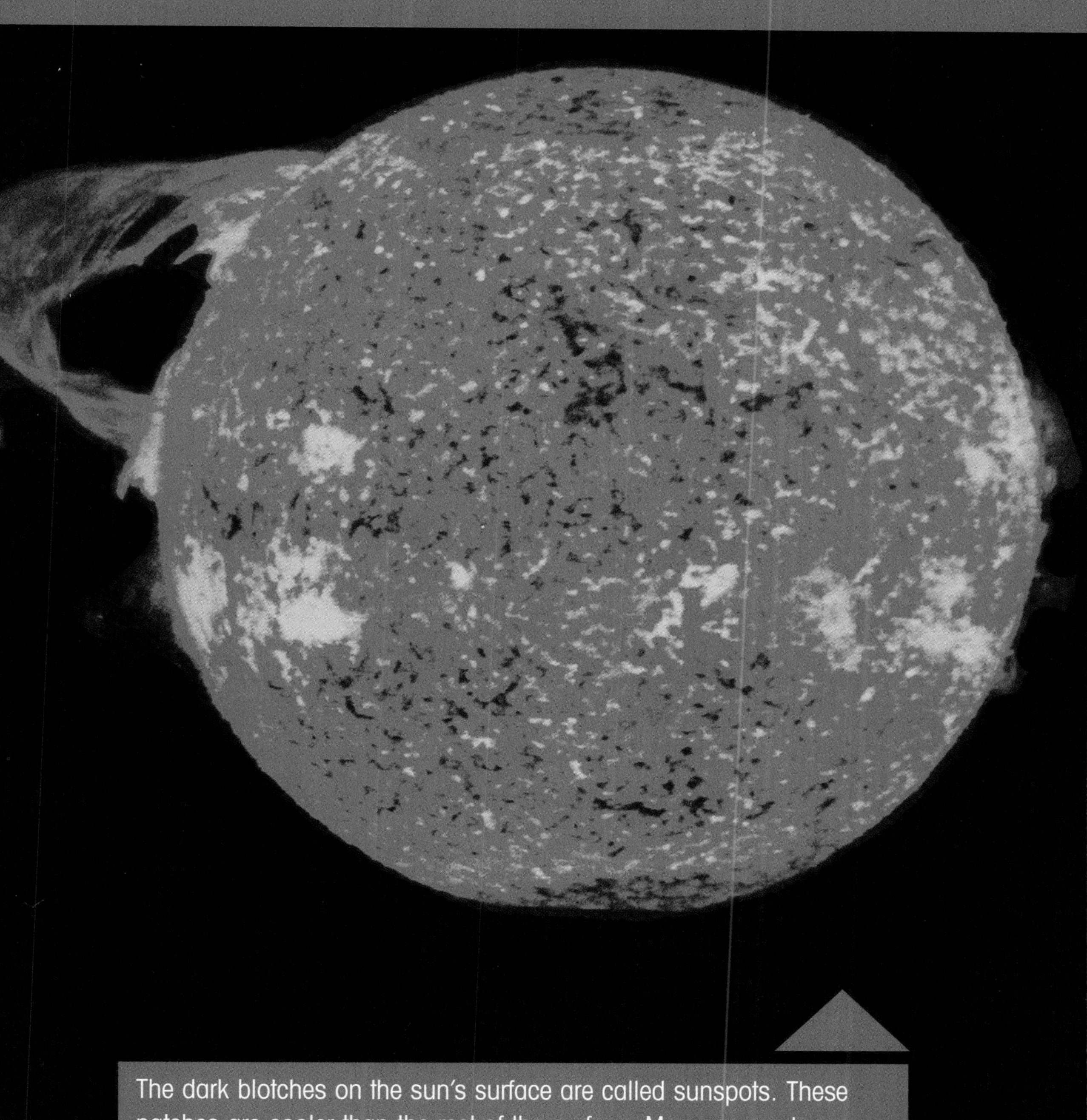

The dark blotches on the sun's surface are called sunspots. These patches are cooler than the rest of the surface. Many sunspots are larger than Earth! The arc of fire in the upper left part of this picture is a solar flare. Solar flares result from explosions on the sun's surface.

The Inner Solar System

Mercury is the closest planet to the sun. It is also one of the smallest planets. Mercury's **diameter**—or the distance across Mercury through its center—is about 3,030 miles. If the sun was the size of a basketball, Mercury would be about the size of a pencil point. Mercury does not have any moons.

The temperatures on Mercury are too extreme for life to exist. Daytime temperatures can reach 800°F. In comparison, the hottest temperature ever recorded on Earth was 134°F. At night, the temperature on Mercury can go down to –275°F. The coldest temperature ever recorded on Earth was –128.6°F.

Mercury orbits the sun in just 88 days, so a year on Mercury is 88 days long. The closest Mercury gets to the sun during its orbit is about 28,600,000 miles. The farthest away it gets is about 43,400,000 miles. What strategy can we use to find how much variation there is in Mercury's distance from the sun during its orbit? That's right, we subtract the smaller number from the larger number.

43,400,000 miles **– 28,600,000 miles** **14,800,000 miles**	**Mercury's distance from the sun varies by 14,800,000 miles during its orbit.**

The surface of Mercury is covered with craters, just like our moon. The craters were caused by meteorites crashing into Mercury's surface. Some of these craters measure more than 250 miles across. The Caloris Basin is the largest crater discovered so far on Mercury. It's about 800 miles across!

Because of the thick clouds around Venus, scientists need special instruments to photograph its surface. The red color shown here is caused by those special instruments.

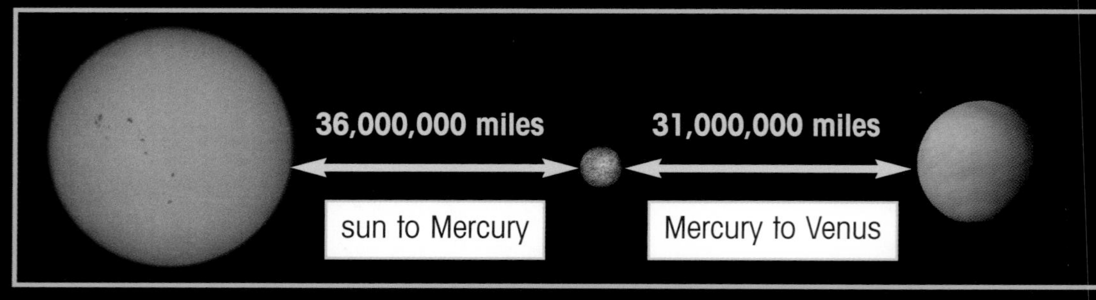

Hot, rocky Venus is the second closest planet to the sun. It's the closest planet to Earth, too. The surface of Venus has mountains, canyons, volcanoes, and plains. The diameter of Venus is about 7,520 miles. If you look in the night sky at the right time of the year, you will be able to see Venus. As far as we know, there is no plant or animal life on Venus. Venus doesn't have any moons, either.

Venus is the hottest of all the planets. The temperature reaches about 890°F. There isn't any water on Venus. The planet does have yellowish clouds, but these are full of acid. The atmosphere on Venus is mostly carbon dioxide, so humans could not breathe there.

Mercury is an average of 36,000,000 miles from the sun. Venus is an average of 31,000,000 miles from Mercury. How far is Venus from the sun? To find the answer, we have to come up with a strategy using the information we have. Which of these 2 operations gives us the answer?

addition

36,000,000 miles
+ 31,000,000 miles
67,000,000 miles

subtraction

36,000,000 miles
– 31,000,000 miles
5,000,000 miles

We can find the distance between the sun and Venus by adding the distance between the sun and Mercury and the distance between Mercury and Venus. The sum is 67,000,000 miles. That is the average distance between the sun and Venus.

Earth is the third planet in our solar system and is about 93,000,000 miles from the sun. Earth is about 7,926 miles in diameter. That's about the same size as Venus. Earth is the only planet we know of that can support life.

Earth **rotates** on its **axis** as it **revolves** around the sun. With all of this spinning, how do we stay put? Gravity keeps our feet on the ground. Gravity is the force between 2 objects that attracts or pulls them toward each other. Gravity is what gives us weight.

Every planet has a different surface gravity. You would weigh more on a planet with a surface gravity stronger than Earth's and less on a planet with a surface gravity weaker than Earth's. The chart on page 13 shows the surface gravity of all the planets in our solar system. We can use this chart to create a strategy to find out how much we would weigh on the surface of another planet. If someone who weighed 100 pounds on Earth traveled to Jupiter, how much would that person weigh? To find out, we need to multiply the person's weight on Earth (100 pounds) by Jupiter's surface gravity (2.6).

100 pounds **x 2.6 Jupiter's surface gravity** **600** **+200** **260.0 pounds**	**Someone who weighs 100 pounds on Earth would weigh 260 pounds on Jupiter. You can check your work by dividing 260 by 2.6 to get the original weight of 100 pounds.**

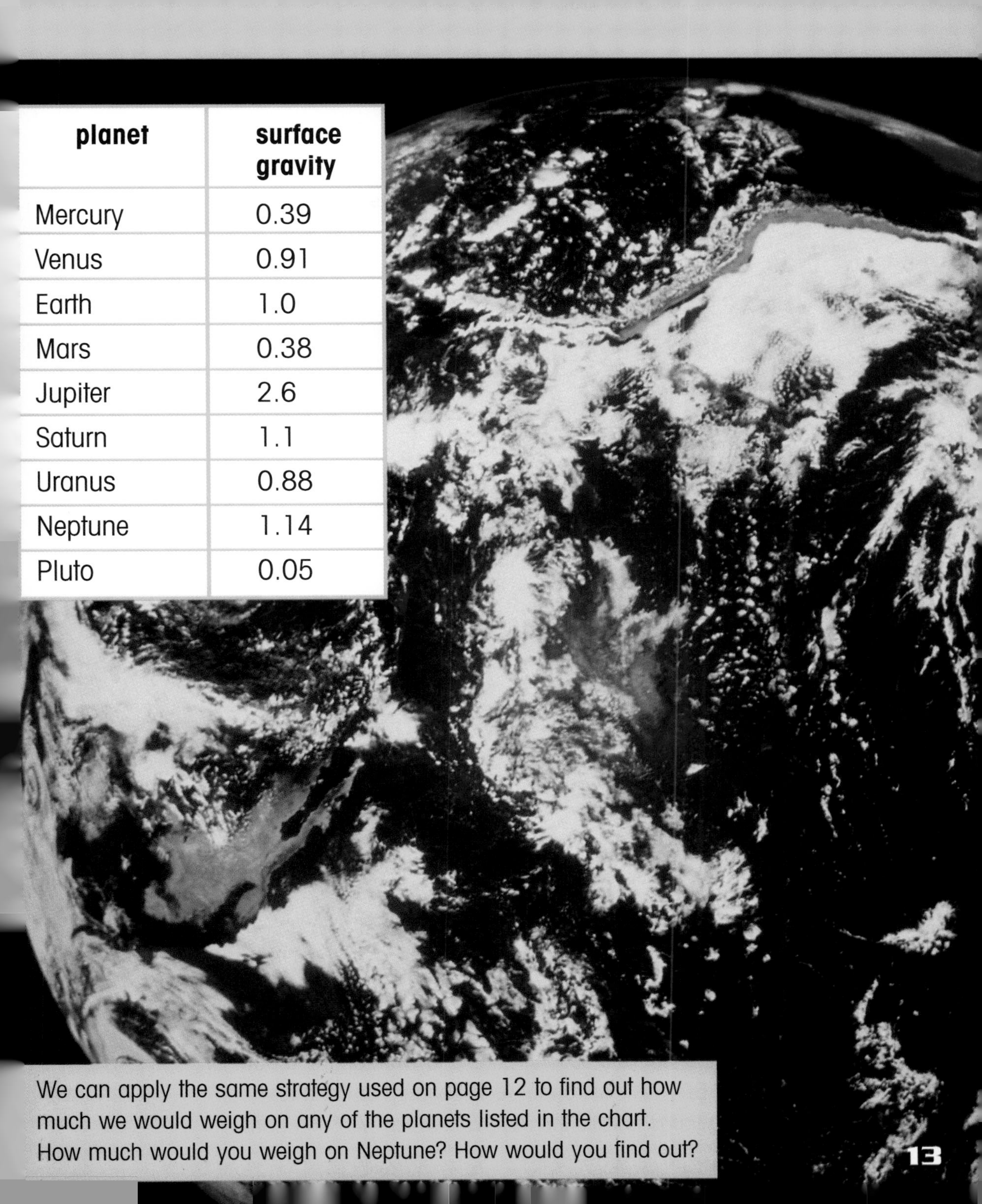

planet	surface gravity
Mercury	0.39
Venus	0.91
Earth	1.0
Mars	0.38
Jupiter	2.6
Saturn	1.1
Uranus	0.88
Neptune	1.14
Pluto	0.05

We can apply the same strategy used on page 12 to find out how much we would weigh on any of the planets listed in the chart. How much would you weigh on Neptune? How would you find out?

On July 16, 1969, astronauts Neil Armstrong and Edwin Aldrin became the first and second people to walk on the moon. On February 6, 1971, astronaut Alan Shepard Jr. hit a golf ball over 200 yards across the moon's surface with one hand!

Earth has only 1 moon. The distance between the moon and Earth is about 238,000 miles. That makes the moon Earth's closest neighbor. The moon is the only space object besides Earth that humans have walked on. From 1969 until 1972, astronauts landed on the moon 6 different times. These missions taught us a lot about the moon. We know the moon's surface is dry and dusty with many craters and mountains. We know that there is no water on the moon. We haven't found any evidence of life there, either.

Since the moon's surface gravity is weaker than Earth's surface gravity, the moon orbits Earth rather than the other way around. Because the gravity on the moon is so much less, walking there is very different. You wouldn't really walk—you'd bounce!

If you weighed 100 pounds on Earth, you would weigh only 16.5 pounds on the moon! What is the surface gravity of the moon? We know that we can find our weight on another planet or moon by multiplying our weight by the surface gravity of that planet or moon. To find out the surface gravity of the moon, we would need to divide our weight on the moon by our weight on Earth.

100	**pounds on Earth**
x ?	**surface gravity of the moon**
16.5	**pounds on the moon**

$$\begin{array}{r} .165 \\ 100\overline{)16.500} \\ -\underline{10\,0} \\ 650 \\ -\underline{600} \\ 500 \\ -\underline{500} \\ 0 \end{array}$$

The surface gravity of the moon is .165.

If the sun were the size of a basketball, the next planet in the solar system—Mars—would be the size of an apple seed. With a diameter of about 4,220 miles, it's a little more than half the size of Earth. We have sent several space probes to look at the atmosphere and the surface of Mars. In January 2004, a probe named *Spirit* landed in an area we call the Gusev Crater. This crater looks a lot like one of Earth's dried-up lake beds. Thanks to the *Spirit* probe, scientists are now certain that Mars once had water on its surface.

The highest volcanic mountain in the solar system is on Mars. It is called Olympus Mons. This mountain is 16 miles high and 340 miles wide. The highest mountain on Earth is Mount Everest in Asia. It's 5.49 miles high. What is the difference in height between these 2 mountains? Subtract to find out. Remember to add a decimal point and 2 zeroes after the 16.

16.00 miles—height of Olympus Mons
– 5.49 miles—height of Mount Everest
10.51 miles difference

Olympus Mons is 10.51 miles higher than Mount Everest.

Mountains are also measured in feet. A mile is equal to 5,280 feet. We can find out how tall a mountain is in feet by using this **formula**: 5,280 x m (number of miles) = f (number of feet). How many feet tall is Olympus Mons?

	5,280	feet per mile
x	16	miles
	31680	
+	5280	
	84,480	feet

Olympus Mons is 84,480 feet high. You can also use this equation to find Mount Everest's height in feet. How many feet taller is Olympus Mons than Mount Everest? What strategy would you use to find out?

The ground on Mars is rocky and red. The red color comes from a mineral called iron. You can often see reddish-colored Mars in the night sky.

The Outer Solar System

Jupiter is the largest planet. Its diameter is 88,823 miles, and its **mass** is more than twice the mass of all the other planets combined. Jupiter's surface is made up of liquid hydrogen and helium. The planet has at least 61 moons and may have more. Ganymede is the largest of Jupiter's moons and is the largest moon in the solar system. It's even larger than Mercury and Pluto! The diameter of Ganymede is 3,280 miles.

In the United States, we measure distances using miles. Other countries measure distances in kilometers. One mile is equal to about 1.61 kilometers. To **convert** from miles to kilometers, multiply the number of miles by 1.61. We call this number a **conversion factor**. Let's use this information to find out how much larger Jupiter's diameter in kilometers is compared to Ganymede's diameter in kilometers.

First, we can make a formula to convert miles to kilometers: m (miles) x 1.61 = k (kilometers). We can use that formula to find out the diameters of Jupiter and Ganymede in kilometers, since we already know that information in miles. Last, we need to subtract Ganymede's diameter in kilometers from Jupiter's diameter in kilometers.

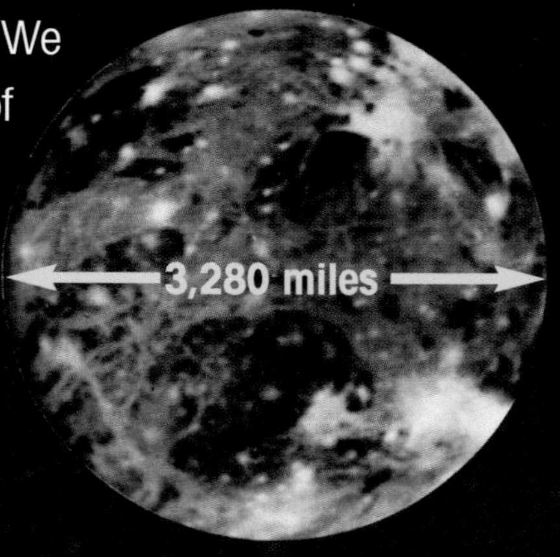

Ganymede

$m \times 1.61 = k$

Step 1: Find Jupiter's diameter in kilometers:

```
   88,823 miles
 x     1.61
 ----------
     88823
   532938
+ 88823
 ----------
 143,005.03 kilometers
```

Step 2: Find Ganymede's diameter in kilometers:

```
   3,280 miles
 x   1.61
 --------
     3280
   19680
+  3280
 --------
 5,280.80 kilometers
```

Step 3: Subtract Ganymede's diameter from Jupiter's diameter:

```
  143,005.03 kilometers
 –  5,280.80 kilometers
 -----------
  137,724.23 kilometers
```

Jupiter's diameter is 137,724.23 kilometers larger than Ganymede's diameter.

Jupiter has stormy weather. One huge storm is called the Great Red Spot. An astronomer first discovered the Great Red Spot over 300 years ago—and it's still stormy today! The Great Red Spot is about 15,000 miles long and 7,500 wide. What are those measurements in kilometers? What strategy would you use to find out?

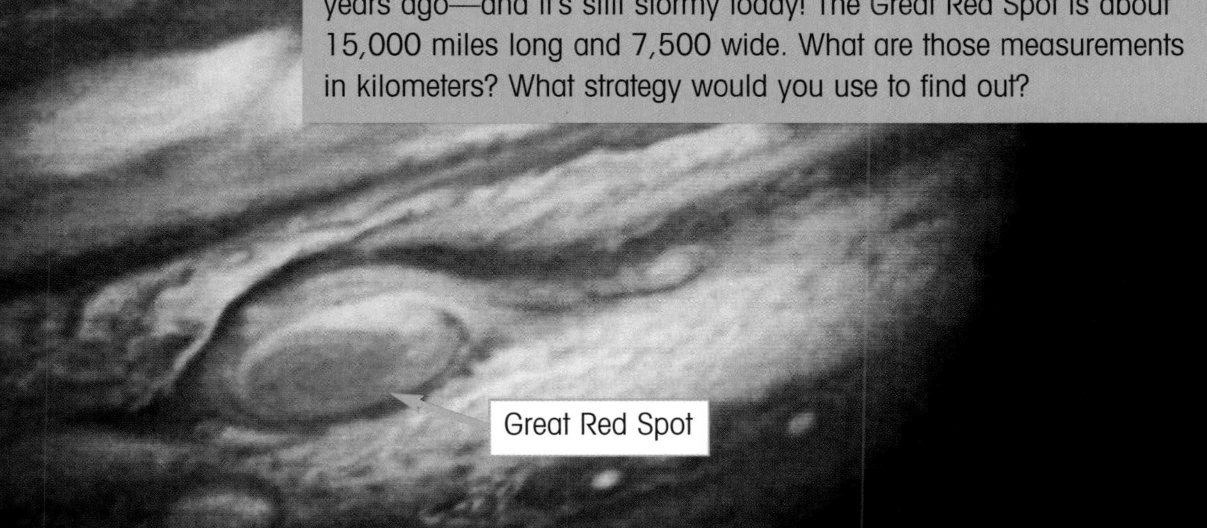

Saturn is the second-largest planet in the solar system. If Jupiter were the size of a quarter, Saturn would be the size of a nickel. The diameter of Saturn is 120,536 kilometers. Like Jupiter, Saturn is made up of liquid hydrogen and helium and has no solid surface. Winds on Saturn reach over 1,700 kilometers per hour. Heat from the planet mixes with the wind to create a yellowish-gold haze in the atmosphere. Scientists discovered that Saturn has 7 large rings, which are made up of many smaller ringlets made of ice.

We have already learned how to convert miles to kilometers. Now let's find out how to convert kilometers to miles. There is .62 mile—or a little over half a mile—in every kilometer, so the formula for converting kilometers to miles is k x .62 = m. Using this formula, let's figure out how much larger Jupiter's diameter in miles is than Saturn's diameter in miles. We know that Jupiter's diameter is 88,823 miles. Find the diameter of Saturn in miles using the formula above. Then subtract Saturn's diameter in miles from Jupiter's diameter in miles.

Step 1: Find the diameter of Saturn in miles:

```
  120,536
 x    .62
 --------
   241072
 + 723216
 --------
 74,732.32
```

The diameter of Saturn is 74,732.32 miles, or about 74,732 miles.

Step 2: Subtract the diameter of Saturn from the diameter of Jupiter:

```
  88,823
- 74,732
--------
  14,091
```

The diameter of Jupiter is about 14,091 miles larger than the diameter of Saturn.

Titan is the largest of Saturn's moons. It's also the second-largest moon in our solar system. The diameter of Titan is 5,150 kilometers. What is the diameter in miles? How much larger is Saturn's diameter in miles than Titan's diameter in miles?

The 3 most distant planets in the solar system are Uranus, Neptune, and Pluto. Uranus is made up of gases and is bluish-green in color. It gets very cold because it is so far from the sun. Temperatures there can drop as low as –300°F to –325°F. Neptune is another bluish-green planet made of gases. There's a dark spot about as big as Earth on Neptune's surface. Scientists think it's a storm like Jupiter's Great Red Spot. Pluto has ice caps at the 2 poles and dark spots at its equator. The distance from Earth to Pluto is so great that Pluto can't be easily seen from Earth. We've learned what we know about Pluto from satellites and a space telescope.

A planet's diameter, as we know, is the distance across the planet through its center. The radius is the distance from the center of a planet to its outer edge. The chart below shows the diameters and radii of the 3 planets farthest from the sun. It is incomplete, but we have enough information to fill in the blanks. If we know the diameter of Neptune, how do we find its radius? If we know the radius of Pluto, how do we find its diameter? Here's a hint: the radius of a circle is $\frac{1}{2}$ its diameter; the diameter of a circle is 2 times its radius.

planet	diameter	radius
Uranus	31,763 miles	15,881.5 miles
Neptune	30,800 miles	
Pluto		715 miles

d (diameter) ÷ 2 = r (radius)	2 x r (radius) = d (diameter)
$$\begin{array}{r} 15{,}400 \text{ miles} \\ 2\overline{)\,30{,}800} \text{ miles} \end{array}$$	$$\begin{array}{r} 715 \text{ miles} \\ \underline{\times\ 2} \\ 1{,}430 \text{ miles} \end{array}$$
The radius of Neptune is 15,400 miles.	The diameter of Pluto is 1,430 miles.

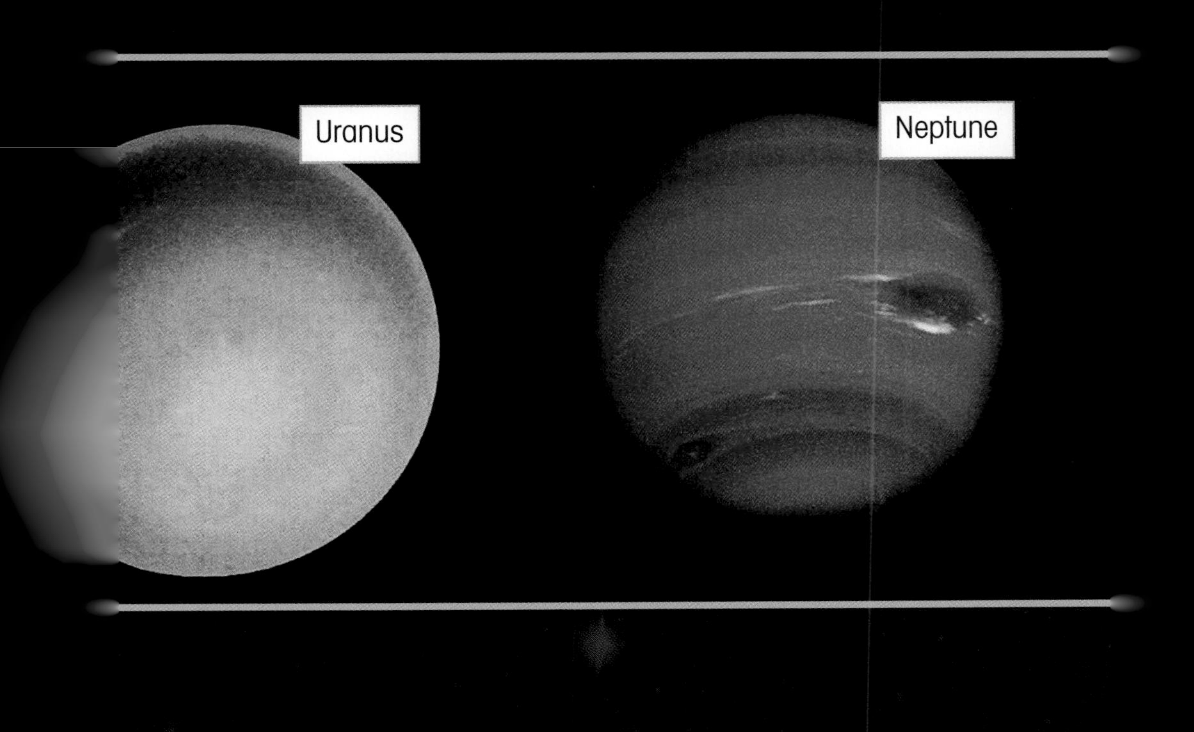

Can you figure out the diameters of Uranus, Neptune, and Pluto in kilometers? What strategy would you use? Look back to pages 18 and 19 for help.

All of the planets in our solar system rotate, or spin. They also revolve around, or orbit, the sun. Each planet revolves and rotates at a different speed, so the length of a day and a year is different on each planet. If you celebrated your birthday on another planet, you'd be a different age than what you are on Earth. How would you find out how old you are on another planet? One way would be to use a ratio.

A ratio is a way to compare 2 values by placing them side by side. We separate the 2 numbers with a colon. For example, if you have 3 green marbles and 6 red marbles, the ratio of green marbles to red marbles is 3:6 (three to six). We can reduce this by dividing both sides by 3 to get 1:2 (one to two). This means that for every green marble, you have 2 red marbles.

planet	Earth days needed to go once around the sun
Mercury	88
Venus	225
Earth	365
Mars	687
Jupiter	4,333
Saturn	10,752
Uranus	30,664
Neptune	60,184
Pluto	90,717

We can use ratios to compare the lengths of years on different planets. The chart on page 24 shows how many Earth days it takes for each of the planets to make 1 complete trip around the sun. For example, 1 year on the planet Mars equals about 687 Earth days. How long is a year on Mars compared to a year on Earth? The ratio of days in a year on Earth compared to the days in a year on Mars is 365:687. To reduce this ratio, divide both sides by 365 so we will have 1 for the first number. Round the answer to the nearest hundredths place.

365:687

```
      1.882
365 ) 687.000
    - 365
      3220
    - 2920
       3000
     - 2920
        800
      - 730
         70
```

1:1.88

One year on Mars is equal to about 1.88 years on Earth. Now, divide your age by 1.88 to find out how old you would be on Mars. Round the answer to the nearest tenth. As an example, let's take someone who is 10 on Earth.

```
        5.31
188 ) 1000.00
     - 940
        600
      - 564
        360
      - 188
        172
```

Don't forget to move the decimal point 2 places to the right in both the divisor and the dividend.

Someone who is 10 on Earth would be about 5.3 years old on Mars.

Small Space Objects

There are other space objects in the solar system besides planets and moons. Comets are small bodies made of ice, rocks, frozen gases, and dust. Sometimes their orbits bring them closer to the sun, and the ice heats up and changes to gas. The gas forms a long, bright tail that can be seen in the night sky. Asteroids are large chunks of metal and rock left over from when the solar system was formed. Most asteroids are between Mars and Jupiter. This area is called the asteroid belt.

English astronomer Edmund Halley observed a comet that appeared in 1682. Halley found evidence of similar comets appearing in the sky in 1531 and 1607. He believed it was the same comet that had been observed by people as far back as 240 B.C. Halley believed that the comet reappeared every 75 to 77 years. He predicted that the comet would return again in 1758, and it did! Halley's Comet still follows this same pattern.

How do you think Halley used this pattern to predict when the comet would reappear? That's right, he added 76 to the date of the last time the comet appeared (1682) to get 1758. He chose 76 because it is the average of 75 and 77, or the number that falls evenly between 75 and 77. Halley's Comet reappeared in 1835, 1910, and 1986. Use Halley's discovery as a strategy to find out when we can expect to see Halley's Comet once again.

Halley's Comet reappears every 75 to 77 years. Between which 2 years can we expect to see Halley's Comet again? To find out when we may see it next, we need to add 75 to 1986 to get a low number, then add 77 to 1986 to get a high number.

year Halley's Comet appeared	time since last appearance
1531	
1607	1607–1531 = 76 years
1682	1682–1607 = 75 years
1758	1758–1682 = 76 years
1835	1835–1758 = 77 years
1910	1910–1835 = 75 years
1986	1986–1910 = 76 years

$$\begin{array}{r} 1986 \\ +\quad 75 \\ \hline ? \end{array}$$

$$\begin{array}{r} 1986 \\ +\quad 77 \\ \hline ? \end{array}$$

Man-Made Space Objects

Since 1957, there have been manned and unmanned missions into space. U.S. astronauts first walked on the moon in 1969. Now, people from the United States and other countries live and work on the International Space Station (ISS). Probes have been sent to every planet except Pluto. Satellites and an outer space telescope called the Hubble Space Telescope send information about the solar system back to Earth. We're learning more about our solar system all the time.

Spacecraft and probes use radio waves to communicate with people on Earth. It takes a while for the messages to get back and forth. Let's say that someone on Earth sends a message to a probe that is traveling to Mars. They do this at 10:00 A.M. The message takes 46 minutes to get to the probe. It takes another 46 minutes to get a message back to Earth. At what time does the sender on Earth get a reply from the probe?

First, add the time for the outbound and return trips.

46	minutes to probe
+ 46	minutes from probe
92	total minutes

Next, change 92 minutes to hours and minutes.

92	total minutes
– 60	minutes in 1 hour
32	minutes more than 1 hour

92 minutes = 1 hour and 32 minutes

Then add 1 hour and 32 minutes to 10:00 A.M.

10:00 A.M.	message leaves
+ 1:32	roundtrip time
11:32 A.M.	message returns

The sender on Earth would receive a reply at 11:32 A.M.

Stardust

The *Stardust* spacecraft journeyed a total of 2 billion miles from Earth to collect dust samples from comets. Scientists think the dust will give us clues about the origin of our solar system, which formed about 4.6 billion years ago.

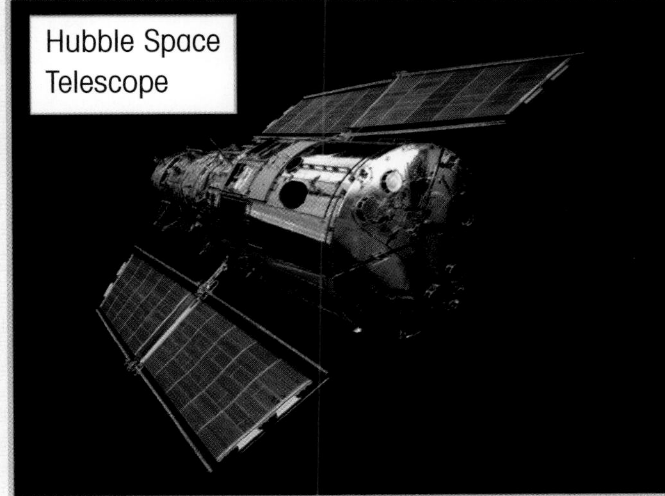

Hubble Space Telescope

International Space Station

Space Strategies

The solar system is a big place, and we are discovering things about it all the time. Without mathematics and problem-solving strategies, astronauts and scientists would not be able to explore space like they can now. You may not become an astronaut or scientist, but that does not mean that you don't need problem-solving strategies, too. People in all kinds of professions need to solve problems every day. Carpenters need to calculate how much wood they will need to finish a job. Doctors need to calculate the proper amounts of medicine for their patients. Chefs need to calculate the amounts of food they will need to feed a specific number of people. All of these people need to use problem-solving strategies.

Problem-solving strategies can also help us out in our everyday lives. For example, let's say you are planning a vacation. You want to travel by car, and you only have 2 weeks to see a total of 10 places. Problem-solving strategies would help you plan your trip. How much gas will you need? How many hours will you need to drive between each place? How much time will you be able to spend at each place? How much money should you plan to take with you? You will need to develop problem-solving strategies to answer these questions before you can take a vacation, or explore the solar system!

Glossary

asteroid (AZ-tuh-royd) A small, rocky space object found especially between Mars and Jupiter.

axis (AK-suhs) The imaginary line around which a planet rotates.

conversion factor (kuhn-VUHR-zhun FAK-tuhr) The number used to convert a measurement in one system to the equal measurement in another system.

convert (kuhn-VUHRT) To change from one form to another.

diameter (dy-A-muh-tuhr) The distance from one side of a planet through the center to the other side of the planet.

formula (FOR-myuh-luh) A general rule or relationship in math, expressed in symbols and often written as an equation.

helium (HEE-lee-uhm) A colorless, odorless gas that is lighter than air.

hydrogen (HY-druh-juhn) A colorless, odorless gas that is lighter than all other elements.

mass (MAS) The amount of matter contained in a planet, moon, or other space object.

nuclear (NOO-klee-uhr) Having to do with reactions that involve the nucleuses of atoms and that release energy.

probe (PROHB) An unmanned spacecraft that gathers information about a planet or region of space and sends the information back to Earth.

revolve (rih-VAHLV) To move in a circular or oval path around the sun.

rotate (ROH-tayt) To turn around an axis.

satellite (SA-tuh-lyt) A man-made object sent into space to orbit Earth, the sun, the moon, or another planet.

strategy (STRA-tuh-jee) A method or plan for solving a problem.

Index

A

asteroid belt, 26
asteroids, 4, 26

C

comet(s), 4, 26
conversion factor, 18
convert, 18, 20

E

Earth, 4, 6, 8, 11, 12, 15, 16, 22, 24, 25, 28

F

formula, 17, 18, 20

G

Ganymede, 18, 19

H

Halley, Edmund, 26
Halley's Comet, 26
Hubble Space Telescope, 28

I

International Space Station, 28

J

Jupiter, 4, 12, 18, 19, 20, 22, 26

M

Mars, 4, 16, 25, 26, 28
Mercury, 4, 8, 11, 18
meteors, 4
moon(s), 4, 8, 11, 15, 18, 26

N

Neptune, 4, 22, 23

P

Pluto, 4, 18, 22, 23, 28
probe(s), 4, 16, 28

R

radius (radii), 22, 23
ratio(s), 24, 25

S

satellite(s), 4, 22, 28
Saturn, 4, 20
Spirit, 16
surface gravity, 12, 15

U

Uranus, 4, 22

V

Venus, 4, 11, 12